Hematology in Traditional Chinese Medicine Cardiology

Hematology in Traditional Chinese Medicine Cardiology

Dr. Anika Niambi Al-Shura, BSc., MSOM, Ph.D
Continuing Education Instructor
Niambi Wellness
Tampa, FL

Medical Illustrator: Samar Sobhy

AMSTERDAM • BOSTON • HEIDELBERG • LONDON
NEW YORK • OXFORD • PARIS • SAN DIEGO
SAN FRANCISCO • SINGAPORE • SYDNEY • TOKYO

Academic Press is an imprint of Elsevier

Academic Press is an imprint of Elsevier
32 Jamestown Road, London NW1 7BY, UK
The Boulevard, Langford Lane, Kidlington, Oxford, OX5 1GB, UK
Radarweg 29, PO Box 211, 1000 AE Amsterdam, The Netherlands
225 Wyman Street, Waltham, MA 02451, USA
525 B Street, Suite 1900, San Diego, CA 92101-4495, USA

Copyright © 2014 Elsevier Inc. All rights reserved.

No part of this publication may be reproduced or transmitted in any form or by any means, electronic or mechanical, including photocopying, recording, or any information storage and retrieval system, without permission in writing from the publisher. Details on how to seek permission, further information about the Publisher's permissions policies and our arrangement with organizations such as the Copyright Clearance Center and the Copyright Licensing Agency, can be found at our website: www.elsevier.com/permissions

This book and the individual contributions contained in it are protected under copyright by the Publisher (other than as may be noted herein).

Notices
Knowledge and best practice in this field are constantly changing. As new research and experience broaden our understanding, changes in research methods, professional practices, or medical treatment may become necessary.

Practitioners and researchers must always rely on their own experience and knowledge in evaluating and using any information, methods, compounds, or experiments described herein. In using such information or methods they should be mindful of their own safety and the safety of others, including parties for whom they have a professional responsibility.

To the fullest extent of the law, neither the Publisher nor the authors, contributors, or editors, assume any liability for any injury and/or damage to persons or property as a matter of products liability, negligence or otherwise, or from any use or operation of any methods, products, instructions, or ideas contained in the material herein.

British Library Cataloguing-in-Publication Data
A catalogue record for this book is available from the British Library

Library of Congress Cataloging-in-Publication Data
A catalog record for this book is available from the Library of Congress

ISBN: 978-0-12-800124-0

For information on all Academic Press publications
visit our website at **store.elsevier.com**

This book has been manufactured using Print On Demand technology. Each copy is produced to order and is limited to black ink. The online version of this book will show color figures where appropriate.

Working together
to grow libraries in
developing countries

www.elsevier.com • www.bookaid.org

DEDICATION

The energy and effort behind the research and writing of this textbook is dedicated to my son, Khaleel Shakeer Ryland. May this inspire and guide you through your journey in your medical studies, career, and life.

ACKNOWLEDGEMENTS

This is a special acknowledgement to my seven-year medical students at Tianjin Medical University (2012−2013) who served as cardiovascular research assistants. May your future medical careers be successful.

An Qi He
Bin Lin Da
Han Jiang
Chen Hua
Jia Ying Luo
Jun Zhang
Lin Lin
Ming Lu
Nang Zhang
Ping Tang
Hu Si Le
Zhao Tian Man
Wen Xing Ning
Xing Wen Zhao
Tang Ying Mei
Li Ying Ying
Xiong Yong Qin
Ding Yu
Li Yan Jun

CONTENTS

Introduction xi

SECTION I BASIC ANATOMY AND PHYSIOLOGY

Chapter 1 Biomedicine Perspective of Blood and Lymph 3
1.1   Part 1: Blood 3
1.2   Part 2: Lymphatic System 5
1.3   Part 3: Genetics 5

Chapter 2 Essence, Qi, and Blood 7
2.1   Part 1: Essence 7
2.2   Part 2: Qi 7
2.3   Part 3: Blood 8

SECTION II RED BLOOD CELLS

Chapter 3 Red Blood Cell Development 13
3.1   Part 1: Red Blood Cell Development 13

Chapter 4 Hemoglobin 15
4.1   Part 1: Protein Subunits 15
4.2   Part 2: The Heme Group 15
4.3   Part 3: The Bohr Effect 16

SECTION III THE INTEGRATIVE IMMUNE SYSTEM

Chapter 5 The Six Stages 21
5.1   Part 1: Purpose of the Six Stages in Cardiovascular Chinese Medicine 21
5.2   Part 2: The Six Stages 22

Chapter 6 The Four Levels 25
6.1   Part 1: Purpose of the Fours Levels in Cardiovascular
        Chinese Medicine..25
6.2   Part 2: The Four Levels..26

Chapter 7 Immune System 1: Leukocytes 29
7.1   Part 1: Types of Leukocytes 29
7.2   Part 2: Leukocyte Disorders 30

Chapter 8 Laboratory Differentials..33
8.1   Part 1: Complete Blood Count 33
8.2   Part 2: Cardiac Markers 34
8.3   Part 3: Lipid Panel..35
8.4   Part 4: Inflammatory Tests...36
8.5   Part 5: Hormone Tests..37

SECTION IV LAB CHARTS

**Chapter 9 Blood Differentiations in Cardiovascular
                   Chinese Medicine 43**
9.1   Differentiations in Cardiovascular Chinese Medicine..................43

INTRODUCTION

The companion course which is required for study with this textbook edition can be found on the Elsevier website and at www.niambiwellness.com.

APPROVING AGENCIES

The course with this textbook is entitled, Integrative Study of Hematology in Cardiology.

This course is approved by the National Certification Commission for Acupuncture and Oriental Medicine (NCCAOM), and is listed as course #1053-007 for 9 PDA points.

This course is approved by the Florida State Board of Acupuncture, and is listed as course # 20-334886 for 10 CEU credits.

COURSE DESCRIPTION

This course briefly covers the basic sciences of blood and immune system from the Western medicine and Traditional Chinese Medicine perspectives. The focus is on laboratory values for diagnosing and monitoring cardiovascular diseases.

COURSE OBJECTIVES

Understand the 5 parts of blood and blood development, genetic blood disorders found in cardiovascular diseases, essence, qi and blood, the immune system including the TCM 6 stages and 4 levels.

Learn the laboratory results and values used for diagnosing and management.

SECTION I

Basic Anatomy and Physiology

CHAPTER 1

Biomedicine Perspective of Blood and Lymph

CHAPTER OBJECTIVES

After studying this chapter, you should be able to:

1. Explain the components of blood.
2. Describe key details about the lymphatic system.
3. Explain the genetic blood disorders and how they affect the cardiovascular system.

1.1 PART 1: BLOOD

Blood carries and delivers oxygen, hormones, and other nutrients to cells and helps maintain the body temperature and the immune system. It also removes waste products and carbon dioxide. The components of blood include plasma, erythrocytes, thrombocytes, leukocytes, and lymphocytes.

1.1.1 Plasma

Plasma is the yellowish clear fluid component which makes up about 55% of blood.

It includes:

1. _____
2. _____
3. _____
4. _____
5. _____
6. _____
7. _____
8. _____
9. _____

Visit the course to find the answers.

1.1.2 Erythrocytes

Erythrocytes are the red blood cells, and they make up about 50% of blood volume, and 3 million cells are continuously produced per second from stem cells located in the bone marrow. Hemoglobin is an iron rich, oxygen attracting protein molecule which creates the red color, and makes up 95% of the erythrocyte. Erythrocytes move and deliver oxygen and other plasma components to all cells, tissues, and organs and carry carbon dioxide and other waste products to the relevant deposit centers of the body. Specific antigens are protein molecules on the surface of erythrocytes which identify the blood type. The antibodies bind to the antigens, and when this happens the process of agglutination occurs which clumps millions of red blood cells together.

10. What is the difference between clumping and clotting?
    _____
    Visit the course to find the answers.

1.1.3 Thrombocytes

Thrombocytes are also called platelets and are cell fragments which circulate within the blood with a life span of about 10 days. The middle of the night and morning hours are time of platelet activation, when they can be a causative factor of serious cardiovascular disease such as in myocardial infarction or stroke. Generally, they are useful in cases of injury and will adhere at the site of a wound or to the walls of damaged vessels to release chemicals which clot and close the wound or vessel. Immediately the 13 different clotting factors are triggered in a cascading sequence.

11. Which factors do hemophiliacs lack?
    _____
    Visit the course to find the answers.

1.1.4 Leukocytes

Leukocytes are white blood cells which make up 0.5–1% of blood, liver, spleen, liver, and lymph nodes. They are produced in bone marrow from stem cells and the thymus gland and have a life span of about 20–36 h.

12. Leukocytes are part of what?
    _____
    Visit the course to find the answers.

1.1.5 Lymphocytes

Lymphocytes are white blood cells which also make up about 0.5–1% of circulating blood. They are produced in the stem cells in bone marrow and function to produce antibodies to pathogenic antigens such as bacteria, viruses, and parasites. This provides immunity to future invasions of the same kind of pathogenic antigen.

13. What is the benefit?
    _____

    Visit the course to find the answers.

1.2 PART 2: LYMPHATIC SYSTEM

The functions of the lymphatic system are to assist the immune system by stimulating an immune response, removal of interstitial fluid from tissues, and transporting of white blood cells. The anatomy of the lymph node includes connective tissue and afferent and efferent compartments called lymph nodules. The afferent vessel allows lymph fluid to enter and flow through the node. The fluid exits the lymph node through the efferent lymphatic vessel, located at the hilum.

14. What is important to know about capillaries?
    _____

    Visit the course to find the answers.

The lymphatic system lacks a pumping mechanism such as in the cardiovascular system, so the movement depends on respiratory movement, skeletal muscle movement, and smooth muscle contraction which is located in vessel walls.

There are three bilateral regions on each side of the body where lymph nodes are located.

15. _____
16. _____
17. _____

    Visit the course to find the answers.

1.3 PART 3: GENETICS

1.3.1 Thalassemia

Thalassemia is a genetic blood disorder in which the body makes an abnormal form of hemoglobin, which destroys erythrocytes causing anemia. There are two main types of thalassemia.

18. What organ failures does thalassemia major cause?
   _____

Visit the course to find the answers.

Thalassemia type	Hemoglobin type	Pain due to other causes
Alpha thalassemia major	Alpha globulin	Middle eastern Chinese South east Asian African
Beta thalassemia minor	Beta globulin	Mediterranean

Copyright © 2014 Anika Niambi Al-Shura. Published by Elsevier Inc. All rights reserved.

1.3.2 Hereditary Hemochromatosis

Hereditary hemochromatosis is a mutation of the C282Y and H63D genes which primarily affects males. It involves an excessive accumulation of iron, which causes damage to the heart, liver, joints, and the testicles.

NOTES

CHAPTER 2

Essence, Qi, and Blood

CHAPTER OBJECTIVES

After studying this chapter, you should be able to:

1. Explain the two types of essence.
2. Describe the origin and action of the six types of qi.
3. Explain the relationship between the zang organs for the production and movement of blood.

2.1 PART 1: ESSENCE

Essence is described in two aspects: the innate and the acquired.

Essence in TCM	
Innate essence	This type of essence is inherited from parents after conception. The lifetime supply is stored in the kidneys after birth to assist with body development.
Acquired essence	This type of essence is derived from regular nutritional intake of food and water.

Copyright © 2014 Anika Niambi Al-Shura. Published by Elsevier Inc. All rights reserved.

1. What are the two noted differences between innate and acquired essence?
   _____
   _____

   Visit the course to find the answers.

2.2 PART 2: QI

Qi represents the basic energy which animates regular life processes and is created partially from acquired essence. It moistens the skin surface to allow antipathogenic qi remain on active duty, helps maintain body temperature, internal blood circulation, and secretion of body fluids.

There are mainly four types of qi: primordial, pectoral, nutritive, and defensive. Two additional types are controlling qi and propelling qi.

Types of Qi	
Primordial qi	This qi is the activated development between puberty and adulthood.
Pectoral qi	This qi is acquired through breathing of fresh air into the lungs and nutritive qi from food.
Nutritive qi	This qi is acquired from food nutrition. It assists in the generation and production of blood.
Defensive qi	This qi is a yang action, also called anti-pathogenic qi. It is created from pectoral qi and nutritive qi.
Controlling qi	Controlling qi allows the opening and closure and dilation and contraction of various pores, sphincters and other orifices throughout the body.
Propelling qi	Propelling qi is the energy which is involved in the physiological functions of the meridians, tissues, organs and organ systems, as well as biochemical mechanisms and products.

Copyright © 2014 Anika Niambi Al-Shura. Published by Elsevier Inc. All rights reserved.

2. Which forms of qi are manifested from exterior factors?
   _____

3. Which forms of qi control internal functions?
   _____

2.3 PART 3: BLOOD

In Chinese medicine, blood is a yin substance. Proper generation and circulation is contingent upon the smoothness of the vessels, the temperature, quality, and the cooperation between the spleen, liver, lung, and heart zang viscera.

Spleen zang	Spleen qi controls blood formation and movement through the vessels.
Liver zang	The liver stores and controls the volume and free movement of blood.
Lung zang	Lung qi mixes fresh air with returning blood to form pectoral qi which is then moved through the heart.
Heart zang	Heart qi propels the blood through the vessels out to the rest of the body.

Copyright © 2014 Anika Niambi Al-Shura. Published by Elsevier Inc. All rights reserved.

NOTES

Module Review Questions
1. Describe the components of blood.
2. Describe key details about the lymphatic system.
3. Explain the genetic blood disorders and how they affect the cardiovascular system.
4. Explain the two types of essence.
5. Describe the origin and action of the six types of qi.
6. Explain the relationship between the zang organs for the production and movement of blood.

Log on at www.niambiwellness.com to access the companion course and quiz for Module 1.

SECTION II

Red Blood Cells

CHAPTER 3

Red Blood Cell Development

CHAPTER OBJECTIVES

After studying this chapter, you should be able to:

1. Explain the process of erythropoeisis.
2. Describe the stages of erythrocyte development.
3. Describe the dimensions of the mature red blood cell and the number count in adults.

3.1 PART 1: RED BLOOD CELL DEVELOPMENT

3.1.1 Erythropoeisis

The cells develop in the bone marrow where erythropoietin (EPO) assists in the formation and later the potential for accumulation of oxygen on its surface.

1. How many days to mature erythocytes circulate the blood?
   _____

   Visit the course to find the answers.

There are eight stages in the development of a red blood cell where gradually through each stage the cell increases in size and volume of cytoplasm.

Erythrocyte development	
1. Hemocytoblast	The early stage of a nucleitic stem cell, which could later become any kind of blood cell.
2. Myeloid	A precursor to a kind of red blood cell.
3. Unipotent stem cell	A cell which is designated to be a certain type of blood cell.
4. Pronormoblast	The precursor stage of an erythrocyte.
5. Erythroblast	The early stage of an immature erythrocyte.
6. Polychromatophilic normoblast	An immature erythrocyte with hemoglobin.
7. Orthochromatic normoblast	The early stage of a reticulocyte, which enters the blood stream.
8. Reticulocyte	An immature circulating red blood cell.

Copyright © 2014 Anika Niambi Al-Shura. Published by Elsevier Inc. All rights reserved.

Hematology in Traditional Chinese Medicine Cardiology. DOI: http://dx.doi.org/10.1016/B978-0-12-800124-0.00003-8
© 2014 Elsevier Inc. All rights reserved.

3.1.2 The Mature Red Blood Cell in the Human Body
- The disk diameter is approximately 6.2−8.2 μm.
- The thickest point is about 2−2.5 μm.
- An adult has about 20−30 trillion erythrocytes.
- Women have about 2._____ erythrocytes per microliter of blood.
- Men have about 5−6 million erythrocytes per microliter of blood.
  Visit the course to find the answers.

NOTES

CHAPTER 4

Hemoglobin

CHAPTER OBJECTIVES

After studying this chapter, you should be able to:

1. Describe the protein subunits of a hemoglobin.
2. Describe the heme group and the role of iron.
3. Describe the Bohr effect in pulmonary and systemic gas exchange.

4.1 PART 1: PROTEIN SUBUNITS

Hemoglobin (Hgb) is the protein in blood which is rich in iron and gives it the red color.

It makes up about 95% of a red blood cell and has an oxygen-binding capacity of 1.34 mL O_2/g, which satisfies the need for the average adult consumption of 250 mL of oxygen per minute. Hemoglobin transports the following gases:

4. _____
5. _____
6. _____
7. _____

Visit the course to find the answers.

Hgb is a globular protein which is structured as an alpha helix or beta helix. Amino acids, also known as histidine, are hydrogen bonded to one of the −NH groups residues or the O in the −CO group.

4.2 PART 2: THE HEME GROUP

Heme groups are the role of iron on hemoglobin. A group contains one iron atom which bonds to the nitrogen molecule of a histidine residue on one end, and on another end bound to one oxygen molecule, allowing a hemoglobin protein to bind to four oxygen molecules.

4.3 PART 3: THE BOHR EFFECT

The Bohr effect is the gas-exchange mechanism in the body. Oxygen which is needed by all cells and tissues in the body remains bound to hemoglobin in the arteries but is released in capillaries. Here, the release of O_2 causes the binding of CO_2 on hemoglobin. In the lungs, the release of CO_2 causes the binding of O_2 on hemoglobin.

8. Who discovered this phenomenon?
   _____

   Visit the course to find the answers.

NOTES

Module Review Questions

1. Explain the process of erythropoeisis.
2. Describe the stages of erythrocyte development.
3. Describe the dimensions of the mature red blood cell and the number count in adults.
4. Describe the protein subunits of a hemoglobin.
5. Describe the heme group and the role of iron.
6. Describe the Bohr effect in pulmonary and systemic gas exchange.

Log on at www.niambiwellness.com to access the companion course and quiz for Module 2.

SECTION III

The Integrative Immune System

CHAPTER 5

The Six Stages

CHAPTER OBJECTIVES

After studying this chapter, you should be able to:

1. Explain the purpose of the pathogenic stages in cardiovascular Chinese medicine.
2. List the pathogenic stages.
3. Describe the symptoms associated with each stage.

5.1 PART 1: PURPOSE OF THE SIX STAGES IN CARDIOVASCULAR CHINESE MEDICINE

The purpose of the six stages in cardiovascular Chinese medicine is to consider the patterns associated with pathogenic invasion on such patients. For example, pathogenic invasions and resulting conditions such as in pneumonia can have a serious effect on the lungs. Complications have the potential to reduce the amount of available oxygen which may lead to cardiac and systemic complications.

1. What is the role of inflammation?
   _____

   Visit the course to find the answers.

5.2 PART 2: THE SIX STAGES

The six stages		
Stage	**Channels**	**Symptoms**
Tai Yang stage Treatment principle: Release the exterior	-Bladder -Small intestine	Fever, chills, aversion to cold and head ache Tongue= thin white coat Pulse= floating, tight
Shao Yang stage Treatment principle: Soothe hyperactivity	-San Jiao -Gallbladder	Alternating hot and cold sensations, headaches, dizziness, blurred vision, hypochondriac pain, irritability, bitter mouth taste and nausea Tongue=thin yellow and white coat Pulse= wiry
Yang Ming stage Treatment principle: Clearing heat	-Stomach -Large intestine	The 5 bigs: fever, sweat, irritability, thirst and pulse Others: aversion to heat and restlessness Tongue= dry yellow coat. Pulse= rapid
Tai Yin Treatment principle: Warm the cold	-Lung -Spleen	Abdominal pain, vomiting, loss of appetite and diarrhea Tongue= sticky white coat Pulse= wiry
Shao Yin Treatment principle: Harmonize, warm	-Heart -Kidneys	Hot above(yang)=fever, insomnia and dry mouth Tongue= red with thin white coat Pulse= rapid Cold below(yin)= cold legs, body aches, frequent clear urination and diarrhea Tongue= pale with white coat Pulse= weak and rapid
Jue Yin (Shock, separation of yin and yang) Treatment principle: balance yin and yang	-Liver -Pericardium	Hunger with inability to eat due to vomiting, alternating sensation of hot and cold, body weakness and possible parasites Tongue= white coat with papillae Pulse= deep and hidden

Copyright © 2014 Anika Niambi Al-Shura. Published by Elsevier Inc. All rights reserved.

NOTES

CHAPTER 6

The Four Levels

CHAPTER OBJECTIVES

After studying this chapter, you should be able to:

1. Explain the purpose of the levels in cardiovascular Chinese medicine.
2. List the pathogenic stages.
3. Describe the symptoms associated with each stage.

6.1 PART 1: PURPOSE OF THE FOURS LEVELS IN CARDIOVASCULAR CHINESE MEDICINE

The purpose of the four levels is similar to the six stages (Chapter 5) in cardiovascular Chinese medicine. It is to consider the patterns associated with pathogenic invasion on such patients. The level penetrated by a pathogenic invasion has a set of associated symptoms, which can guide strategies in treatment. For example, inflammation is part of the innate immune response to certain pathogens. Chronic states can trigger the buildup of cholesterol in the vessels over time, and can also contribute to a rise in blood pressure, myocardial infarction, and stroke.

2. At which level might be associated with inflammatory response triggering emergency cardiovascular symptoms? _____

   Visit the course to find the answers.

6.2 PART 2: THE FOUR LEVELS

The four levels		
Level	**Syndrome types**	**Symptoms**
Wei level Tai yang stage Exterior heat	Wind heat Dry heat Summer heat Damp heat	Aversion to wind, fever, chills, headache, sweating, thirst, sore throat, dry mouth with thirst Tongue= thin white coat Pulse= floating and tight
Qi level Yang ming stage Interior heat	Lung and stomach heat Intestinal heat Spleen damp heat	The 5 "bigs": fever, sweat, irritability, thirst and pulse Others: aversion to heat and restlessness Tongue= dry yellow coat Pulse= rapid
Ying level Nutritional stage Heat and fire	Heat attack of the heart Heat attacking the pericardium	Irritability, mental/emotional disturbances, night fever, insomnia, dry mouth and thirst Tongue= red Pulse= rapid
Xue level Attack on blood	Liver and chong heat Severe heat and wind Severe heart heat Severe blood heat	Irritability, alternating emotional states, dry skin, rashes or twitching, epistaxis and various bleeding disorders Tongue= red Pulse= rapid and wiry

Copyright © 2014 Anika Niambi Al-Shura. Published by Elsevier Inc. All rights reserved.

NOTES

CHAPTER 7

Immune System 1: Leukocytes

CHAPTER OBJECTIVES

After studying this chapter, you should be able to:

1. Explain the purpose of leukocyte study in cardiovascular Chinese medicine.
2. Describe the types of leukocytes.
3. Describe leukocyte disorders.

Leukocytes are also called white blood cells, which are part of the immune system and signal the presence of a pathogen. The purpose of the study of leukocytes in cardiovascular Chinese medicine is to identify the conditions in the body for inflammation and pathogenic invasion during the six stages and four levels, which can be used as a strategy for treatment using Chinese medicine or integration of Chinese and Western medicine.

7.1 PART 1: TYPES OF LEUKOCYTES

Leukocytes defend the body against pathogenic invasions. An increase in the number is called leukocytosis, and a decrease is called leukopenia. Leukocytes are identified as being either granulocytes or agranulocytes.

- Granulocytes: 3._____
- Agranulocytes: 4._____
  Visit the course to find the answers.

Types of white blood cells.

Leukocyte	Type	Description
Neutrophil	Granulocyte	-Defends against bacterial or fungal infection. -Most commonly released with early acute inflammation. -Makes up 60% of total leukocyte count. -The life span is about 5 days.
Eosinophil	Granulocyte	-These are inflammatory cells. -These respond in allergic reactions and parasitic invasions.
Basophil	Granulocyte	-These are responsible for allergic reactions. -These release histamine which causes vasodilation.
Monocyte	Agranulocytes	-These are known as macrophages which are responsible for phagocytosis of pathogenic cells. -They present pathogens to T cells which kill them.

Copyright © 2014 Anika Niambi Al-Shura. Published by Elsevier Inc. All rights reserved.

7.2 PART 2: LEUKOCYTE DISORDERS

7.2.1 Leukocyte Increase
Leukocytosis: 5._____

7.2.2 Leukocyte Decrease
Leukopenia: 6._____

Visit the course to find the answers.

Leukocyte disorders			
Leukocytosis		**Leukopenia**	
Neutrophilia	-These increase in count. -Caused by primary disease types.	Neutropenia	-Decreases in count. -Are often drug induced.
Eosinophilia	-Presence is not normal. -Incidence may happen without cause. -May be seen in the blood with allergies, morbid cholesterol levels, and interstitial nephropathy.	Lymphocytopenia	-Decreases in count. -Incidence is seen in immune compromised patients. -Often drug induced -May be due to zinc deficiencies.

Copyright © 2014 Anika Niambi Al-Shura. Published by Elsevier Inc. All rights reserved.

NOTES

Module Review Questions

1. Explain the purpose of the pathogenic stages in cardiovascular Chinese medicine.
2. List the pathogenic stages.
3. Describe the symptoms associated with each stage.
4. Explain the purpose of the levels in cardiovascular Chinese medicine.
5. List the pathogenic stages.
6. Describe the symptoms associated with each stage.
7. Explain the purpose of leukocyte study in cardiovascular Chinese medicine.
8. Describe the types of leukocytes.
9. Describe leukocyte disorders.

Log on at www.niambiwellness.com to access the companion course and quiz for Module 3.

CHAPTER 8

Laboratory Differentials

CHAPTER OBJECTIVES

After studying this chapter, you should be able to:

1. Explain the significance of the complete blood count (CBC) in cardiovascular disease screening, diagnosing, and monitoring.
2. Explain the significance of the cardiac markers for screening, diagnosing, and monitoring.
3. Explain the significance of the lipid panel in cardiovascular disease screening, diagnosing, and monitoring.
4. Explain the significance of the inflammatory tests in cardiovascular disease screening, diagnosing, and monitoring.
5. Explain the significance of the hormone tests in cardiovascular disease screening, diagnosing, and monitoring.

8.1 PART 1: COMPLETE BLOOD COUNT

The purpose of the CBC is to determine the cause of complaints about fatigue, weakness, determine the amount of blood loss during a hemorrhage, determine if there is an infection or inflammation present, view the affects of certain medications or herbal formulas on the blood or immune system, and diagnose polycythemia or anemia.

Test	Reference interval	Interpretation
WBC Per cubic mL	5,000–10,000	Indicates the presence of infection or inflammation.
RBC Millions/mcL	Men: 4.4–5.5 Women: 4.0–5.9	–Low count indicates a hemorrhage, anemia or iron deficiency –High count indicates clotting as in atherosclerosis.
Hemoglobin (HgB) g/L	Men: 140–174 Women: 120–160	–Measures the ability of RBC to carry oxygen. –A good test for anemia.
Hematocrit (HCT)	Men: .42–.52% Women: .36–.48%	–Measures the percent of erythrocytes in the blood. –A good test for anemia.
Mean corpuscular volume (MCV) cubic mm	82–98	–Size of the red blood cells.
Mean corpuscular hemoglobin (MCH) cubic mm	26–34	–Amount of hemoglobin in RBCs.
Red cell distribution width (RDW)	11.5–14.5%	–Compares the size and shapes of all RBCs.
Platelets	–Low count indicates idiopathic thrombocytopenic purpura and splenomegaly.	
Mean platelet volume (MPV)	Range: 9.7–12.8/ femoliter (fL) Normal: 7–11	–Measures the average size of the platelets.
Neutrophils	50%	These tests are used to measure the presence of specific infections and inflammation.
Basophils	0–1%	
Eosinophils	0–3%	
Neutrophils	3–6%	
Lymphocytes	25–40%	
Monocytes	3–7%	
Reticulocytes	Measures the count of new red blood cells, which may indicate that mature ones are depleting fast due to destruction or a hemorrhage.	
Erythrocyte sedimentation rate (ESR)	Measures the rate erythrocytes settle inside of the test tube. The cells settle slowly with the presence of inflammation.	

Copyright © 2014 Anika Niambi Al-Shura. Published by Elsevier Inc. All rights reserved.

8.2 PART 2: CARDIAC MARKERS

Cardiac biomarkers are used to detect emergency acute cardiac events and to monitor while under hospital observation.

Cardiac biomarkers

Marker	Description	Indication period
Troponin I and T	-Acute myocardial infarction -Heart failure -Myocarditis	-Elevates in 2-4 hours -Peaks in 12 hours -Must be measured within 3 days -May persist for 7 days
Creatinine kinase (CK-MB)	-Diagnosis of acute myocardial infarction	-Releases immediately -Peaks within 24 hours -Normal in 2 days
Pro-brain natriuretic peptide	-Heart failure indication -Poor prognosis value= >100 -Better prognosis value= < 80	-Time frame is ongoing
Glycogen phosphorylase isoenzyme BB	-Detects ischemia and MI -Early indicator in acute coronary syndrome -Detects unstable angina	-Elevates in 1-3 hours -Persists up to 7 hours

Copyright © 2014 Anika Niambi Al-Shura. Published by Elsevier Inc. All rights reserved.

8.3 PART 3: LIPID PANEL

The lipid panel provides information on total cholesterol (TC), low-density lipoprotein (LDL), high-density lipoprotein (HDL), and triglycerides. These tests help determine the levels of certain types of lipids. Sometimes, the glucose test is included which helps to monitor the blood sugar levels.

1. What is the purpose of the tests?
   _____
   Visit the course to find the answers.

Lipid panel

Glucose	Range: 65-99 mg/dL Normal: 70-85 mg/dL
Cholesterol	Range: 100-199 mg/dL Normal: 180-200 mg/dL
LDL	Range: 0-99 mg/dL Normal: Under 100 mg/dL
HDL	Range: 40-59 mg/dL Normal: Over 55 mg/dL
Triglycerides	Range: 0-149 mg/dL Normal: Under 100 mg/dL
Hemoglobin	Range: 4.5-5.7% Normal: 4.5%

Copyright © 2014 Anika Niambi Al-Shura. Published by Elsevier Inc. All rights reserved.

8.4 PART 4: INFLAMMATORY TESTS

8.4.1 Fibrinogen
High fibrinogen levels are markers for inflammation.

2. What are the risk factors?
   _____
   Visit the course to find the answers.

8.4.2 Homocysteine
Homocysteine is an amino acid which is formed during methionine metabolism.

3. What is the risk factor?
   _____
   Visit the course to find the answers.

8.4.3 C-reactive Protein
C-reactive protein is a marker of systemic inflammation, which is a risk factor for cardiovascular diseases such as hypertension, coronary artery disease, diabetes mellitus, and metabolic syndrome.

Inflammation tests	
Fibrinogen	Range: 193–423 mg/dL Normal: 295–369 mg/dL
Homocysteine	Men Range: 4.3–15.3 µmol/L Normal: < 7.2 µmol/L
	Women Range: 3.3–11.6 µmol/L Normal: < 7.2 µmol/L
C-reactive protein	Men Range: 0–3 mg/L Normal: <0.55 mg/L
	Women Range: 0–3 mg/L Normal: <1.5 mg/L

Copyright © 2014 Anika Niambi Al-Shura. Published by Elsevier Inc. All rights reserved.

8.5 PART 5: HORMONE TESTS

8.5.1 Dehydroepiandrosterone

Dehydroepiandrosterone (DHEA) is produced by the adrenal glands, and is a precursor to male and female reproductive hormones.

7. What do balanced levels indicate?
   _____
   Visit the course to find the answers.

8.5.2 Thyroid Stimulating Hormone

Thyroid stimulating hormone (TSH) is produced in the thyroid.

8. What are the pathological symptoms?
   _____
   Visit the course to find the answers.

Hypothyroidism	Hyperthyroidism
• hypercholesterolemia	• atrial fibrillation
• weakness and fatigue	• heart palpitations
• low grade depression	• excessive weight loss
• cold intolerance	• nervousness
• weight gain	• insomnia
• dry skin	• heat intolerance
• constipation	• weakness and fatigue

Copyright © 2014 Anika Niambi Al-Shura. Published by Elsevier Inc. All rights reserved.

8.5.3 Estrogen

Estrogen is a class of hormones which also includes estriol, estradiol, and estrone. It is a female hormone which is present in both men and women but in much higher levels in women.

9. What risk cardiovascular risk factor may be reduced?
   _____
   Visit the course to find the answers.

8.5.4 Testosterone (Free)

Testosterone is produced in the adrenal glands in men and women. It is produced in higher levels within the testes of men, and in lower levels within the ovaries in women.

10. What problems do low levels cause in men?
   _____

11. What problems do low levels cause in women?
   _____

Visit the course to find the answers.

Hormone tests	
TSH	Range: 0.35–5.50 mU/L Normal: 0.35–2.1 mU/L
DHEA	Men Range: 280–640 µg/dL Normal: 400–500 µg/dL
	Women Range: 65–380 µg/dL Normal: 350–430 µg/dL
Estrogen	Men Normal: 10–50 picograms(pg)/mL
	Women Pre-menopausal: 30–300 pg/mL Post menopausal: 0– 30 pg/mL
Testosterone	Men Range: 6.6–26.5 pg/mL Normal: 15–26.5 pg/mL
	Women Range: 0–2.2 pg/mL Normal: 1.4–2.2 pg/mL

Copyright © 2014 Anika Niambi Al-Shura. Published by Elsevier Inc. All rights reserved.

NOTES

Log on at www.niambiwellness.com to access the companion course and quiz for Module 4.

SECTION IV

Lab Charts

CHAPTER 9

Blood Differentiations in Cardiovascular Chinese Medicine

CHAPTER OBJECTIVES

After studying this chapter, you should be able to:

1. List the common symptoms of blood stagnation.
2. List the cardiovascular symptoms of blood stagnation.
3. List the common symptoms of blood stasis.
4. List the cardiovascular symptoms of blood stasis.
5. List the common symptoms of blood deficiency.
6. List the cardiovascular symptoms of blood deficiency.

9.1 DIFFERENTIATIONS IN CARDIOVASCULAR CHINESE MEDICINE

\multicolumn{4}{c}{TCM blood pattern diagnosing in cardiology}
Differentiation	Symptoms	Tongue	Pulse
Blood stagnation	Fixed, stabbing pain and clotting. Cardiovascular: dyspnea and Orthopnea, hemorrhage, palpitations, lipid imbalance and blood pressure elevations.	Purple	-Choppy -Wiry (pain)
Blood stasis	Cyanosis, edema, dark colored skin. Cardiovascular: infarction, ischemia, hemorrhage and chest pain.	Dark purple	-Wiry - Knotted -Intermittent
Blood deficiency	Pale lips and complexion, cold body sensation, blurred vision, insomnia, dizziness, numbness or tremors, dry skin and hair and constipation. Cardiovascular: anemia, inflammation, blood pressure elevations, dyspnea and orthopnea.	Pale	-Rapid -Weak

Copyright © 2014 Anika Niambi Al-Shura. Published by Elsevier Inc. All rights reserved.

NOTES

Course Review Questions
1. Describe the components of blood.
2. Describe key details about the lymphatic system.
3. Explain the genetic blood disorders and how they affect the cardiovascular system.
4. Explain the two types of essence.
5. Describe the origin and action of the six types of qi.
6. Explain the relationship between the zang organs for the production and movement of blood.
7. Explain the process of erythropoeisis.
8. Describe the stages of erythrocyte development.
9. Describe the dimensions of the mature red blood cell and the number count in adults.
10. Describe the protein subunits of a hemoglobin.
11. Describe the heme group and the role of iron.
12. Describe the Bohr effect in pulmonary and systemic gas exchange.

13. Explain the purpose of the pathogenic stages in cardiovascular Chinese medicine.
14. List the pathogenic stages.
15. Describe the symptoms associated with each stage.
16. Explain the purpose of the levels in cardiovascular Chinese medicine.
17. List the pathogenic stages.
18. Describe the symptoms associated with each stage.
19. Explain the purpose of leukocyte study in cardiovascular Chinese medicine.
20. Describe the types of leukocytes.
21. Describe leukocyte disorders.
22. Explain the significance of the complete blood count in cardiovascular disease screening, diagnosing, and monitoring.
23. Explain the significance of the cardiac markers for screening, diagnosing, and monitoring.
24. Explain the significance of the lipid panel in cardiovascular disease screening, diagnosing, and monitoring.
25. Explain the significance of the inflammatory tests in cardiovascular disease screening, diagnosing, and monitoring.
26. Explain the significance of the hormone tests in cardiovascular disease screening, diagnosing, and monitoring.
27. List the common symptoms of blood stagnation.
28. List the cardiovascular symptoms of blood stagnation.
29. List the common symptoms of blood stasis.
30. List the cardiovascular symptoms of blood stasis.
31. List the common symptoms of blood deficiency.
32. List the cardiovascular symptoms of blood deficiency.

This also concludes the integrative anatomy and pathophysiology in TCM cardiology course. It is strongly suggested that you log onto the courses at the companion web sites to review the course modules. Next, submit course documents and complete the final exam.

Upon passing the exam, you will receive completion certificates which include your name and practice license number, along with the specific number of credit hours awarded for this course. Electronic transmission of CEU and PDA credits will be sent to NCCAOM and your state medical board.

Printed in Poland
by Amazon Fulfillment
Poland Sp. z o.o., Wrocław